Dinosaur Discovery

MOSASAURUS

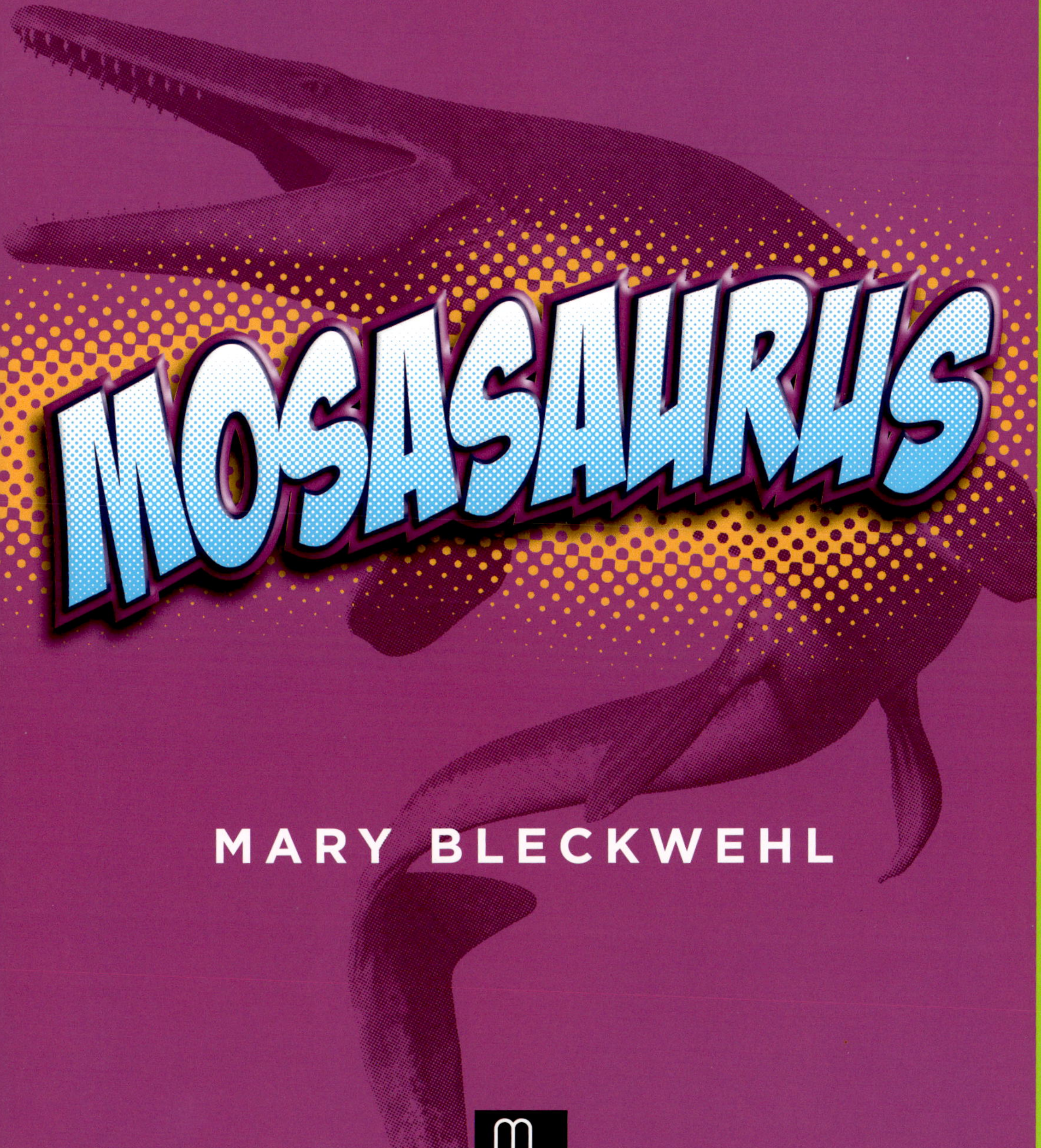

MARY BLECKWEHL

BLACK RABBIT BOOKS

Bolt is published by Black Rabbit Books
P.O. Box 227, Mankato, Minnesota, 56002.
www.blackrabbitbooks.com

Alissa Thielges, editor; Rhea Magaro, designer and photo researcher

Library of Congress Cataloging-in-Publication Data
Names: Bleckwehl, Mary E., author.
Title: Mosasaurus / by Mary Bleckwehl.
Description: Mankato, Minnesota: Black Rabbit Books, [2024] | Series: Dinosaur discovery | Includes bibliographical references and index. | Audience: Ages 8–12 | Audience: Grades 4–6 | Summary: "With strong flippers and shark-like teeth, Mosasaurus once ruled the oceans. Reluctant readers discover what life was like for this Cretaceous sea reptile through diagrams, graphs, powerful illustrations, and fun text"—Provided by publisher.
Identifiers: LCCN 2023035688 (print) | LCCN 2023035689 (ebook) | ISBN 9781623109097 (library binding) | ISBN 9781623109257 (ebook)
Subjects: LCSH: Mosasaurus—Juvenile literature.
Classification: LCC QE862.S65 B54 2024 (print) | LCC QE862.S65 (ebook) | DDC 567.9/37—dc23/eng/20231020
LC record available at https://lccn.loc.gov/2023035688
LC ebook record available at https://lccn.loc.gov/2023035689

Printed in China.

Image Credits

Alamy: dotted zebra 14, Mohamad Haghani 6–7, Stocktrek Images, Inc. cover; iStock: CoreyFord 32, MR1805 18 (bkgd), Racksuz 8, ZU_09 13 (main); public domain 17, 22, 24–25; Science Source: Arthur Dorety 4–5, Mark Garlick 21 (main); Shutterstock: Andrey_Kuzmin 3, BearFotos 21 (clams), Buch and Bee 25, Dotted Yeti 1, Dotted Yeti 28–29, mognev 9 (Mosasaurus), N. Rotteveel 23, Ryan M. Bolton 29, SciePro 10–11, SciePro 31, Sergey Uryadnikov 26–27, Starostov 28, Yarr65 13 (now), Zhenyakot 9 (shark)
Every effort has been made to contact copyright holders for material reproduced in this book. Any omissions will be rectified in subsequent printings if notice is given to the publisher.

CONTENTS

CHAPTER 1

Meet the MOSASAURUS

A big lizard-like creature swims near the shore. It is a young, hungry Mosasaurus. On the prowl, it spies a turtle. It's an easy **prey** to snatch up. But the **reptile** is still hungry. A fish catches its eye. The Mosasaurus opens its jaw wide. *Chomp*. It gulps the fish down in one bite.

Top of the Food Chain

After its meal, the Mosasaurus floats out to sea. Suddenly, a dark shadow lunges at it. It's another Mosasaurus! This one is as long as a school bus. It tries to sink its spiky teeth into the smaller reptile's thick neck. But the younger Mosasaurus is tough. It quickly swims off to safety

Mosasaurus means "lizard of the Meuse River." This river is where its fossils were first found.

T-Rex of the Sea

Mosasaurus was not a dinosaur. It was a sea reptile and one of the biggest **species** of the mosasaur family. Most were the size of a dolphin. But some grew longer than Tyrannosaurus rexes.

COMPARING SIZES

This reptile could swallow animals whole. Few prey escaped its massive jaw. Its 40 to 50 teeth were set in two rows. Mosasaurus also moved fast! Paddle-like limbs and a strong tail steered it through the water.

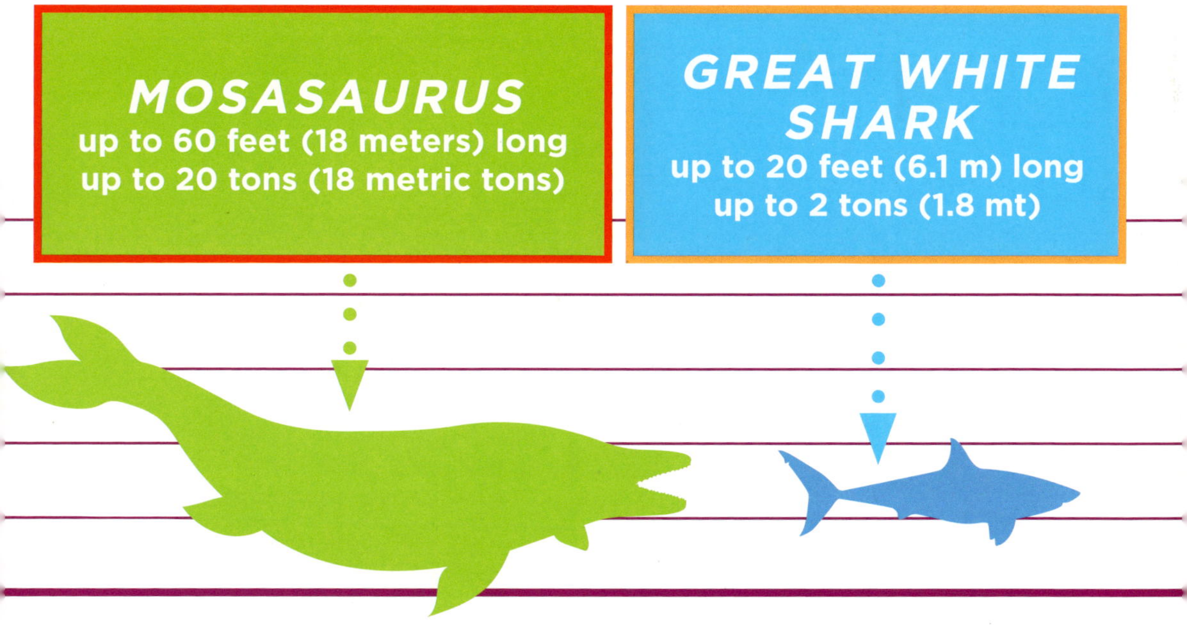

MOSASAURUS FEATURES

DIAMOND SHAPED SCALES
SHARK-LIKE TAIL
STUMPY PADDLE-LIKE FLIPPERS

CHAPTER 2

Where and WHEN THEY LIVED

The Mosasaurus ruled the oceans about 66 to 82 million years ago. This was the late Cretaceous period. Oceans covered a lot of land in that time. Sea water used to split North America in two. Mosasaurus lived in that seaway.

North America Then and Now

69–75 million years ago

now

This reptile had no gills. Like whales, it likely swam near the surface to breathe air.

A Different World

The world was much warmer back then. The poles had no ice. It's possible that Mosasaurus swam all around the world. It **adapted** to both salt and fresh waters. This gave the sea reptile many food choices.

First Fossils

Mosasaurus bones were first found in 1764. They were in a Netherlands **quarry**. More fossils have been found in North America. They show scientists where this sea creature lived. Many teeth have been found. They were sharp and as long as 3 inches (7.6 cm).

SOUTH DAKOTA

A skeleton was once found in South Dakota. It had five unborn babies inside. This means Mosasaurus likely gave birth to live young.

WHERE MOSASAURUS FOSSILS HAVE BEEN FOUND

ASIA
EUROPE
AFRICA

CHAPTER 3

WHAT THEY ATE

Mosasaurus ate a lot! It spent much of its time hunting. As a **carnivore**, it wasn't picky. It ate anything that moved. Birds, fish, **mollusks**, and squid were all on the menu. So were smaller Mosasaurs. They even snatched smaller dinosaurs that were close to shore.

mollusks

The movie *Jurassic World* includes a Mosasaurus. The sea reptile leaps straight out of the water. But the real creatures were not that big.

How They Ate

Scientists believe this giant prowled coastlines and open water. Its sense of smell wasn't very good. But fossils show it had big eyes. It could probably see well under the water.

Mosasaurus swallowed small prey in one gulp. For bigger food, it grabbed the prey's snout. Sharp teeth tore chunks off to gulp down.

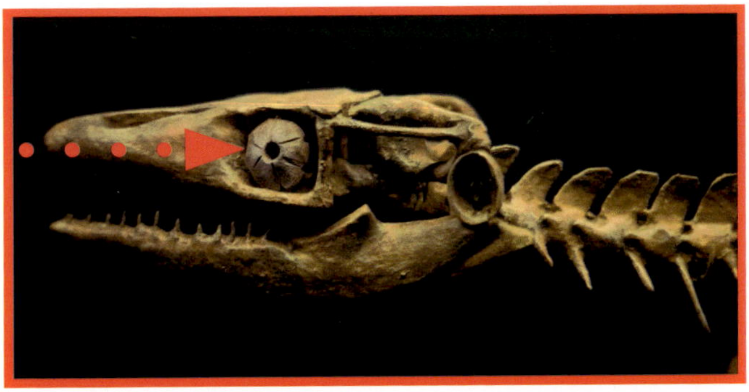

CHAPTER 4

New DISCOVERIES

Paleontologists still study this creature. In 2022, a long fossil was found in north Texas. The find may help us know how this sea reptile **evolved**. Some scientists think it began as a land lizard and adapted to ocean life. Others think it used to be a snake that grew flippers. The answer is still unknown. We need more fossils to know the truth.

How Mosasaurs May Have Evolved

MILLION YEARS AGO

JURASSIC AGE
about 199 to 145 million years ago

END OF JURASSIC PERIOD
145 million years ago

MONITOR LIZARDS LIVE ON LAND
145-100 million years ago

CRETACEOUS AGE
about 145 to 65 million years ago

LIZARDS BEGAN HUNTING IN COASTAL WATERS
100-90 million years ago

MOSASAURS LIVE IN THE OCEANS FULL-TIME
90-66 million years ago

Learning about the Past

Mosasaurus disappeared about the same time dinosaurs did. But how quickly did the sea creatures die? It is unclear. The sea levels likely dropped as the temperatures got cooler. Food was hard to find. At some point, the big sea reptiles likely starved and died. Knowing how this happened could help us know how climate change affects sea life today.

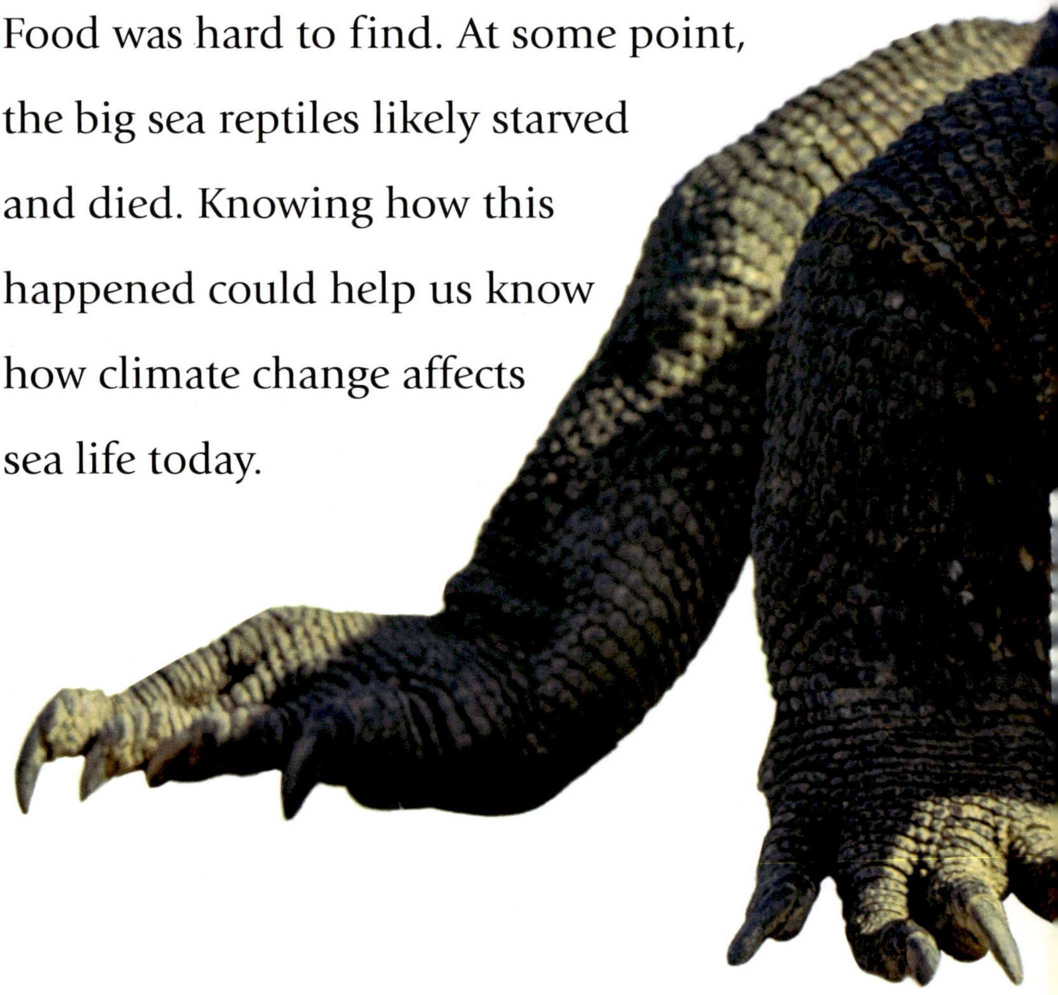

The Komodo dragon is related to Mosasaurus.

BY THE NUMBERS

20 YEARS
ESTIMATED LIFE SPAN

3.5 elephants
EQUAL WEIGHT OF MOSASAURUS

100
NUMBER OF VERTEBRAE IN ITS BACKBONE

3
INCHES
(7.6 CM)
LONGEST TOOTH FOSSIL

56
feet
(17 m)
LONGEST MOSASAURUS FOSSIL

GLOSSARY

adapt (uh-DAPT)—to change something so it is better suited for a purpose

carnivore (CAR-nuh-vohr)—an animal that eats meat

evolve (ih-VOLV)—to slowly change

mollusk (MOL-uhsk)—a group of animals that do not have a backbone and usually live in a shell for protection

paleontologist (pay-lee-ohn-TALL-uh-jist)—a scientist who studies fossils

prey (PRAY)—an animal hunted or killed for food

reptile (REP-tahyl)—a cold-blooded animal that breathes air and has a backbone; most reptiles lay eggs and have scaly skin.

species (SPEE-seez)—a class of individuals that have common characteristics and share a common name

LEARN MORE

BOOKS

Hansen, Grace. *Mosasaurus.* Mankato, Minn.: Abdo Kids Jumbo, 2020.

Naish, Darren. *Ancient Sea Reptiles.* Washington, D.C.: Smithsonian Books, 2023.

Nelson, Jake. *I'm a Mosasaurus.* Ann Arbor, Mich.: Cherry Lake Publishing, 2020.

WEBSITES

Dinosaur Discoveries: Mosasaurus
pbskids.org/video/dinosaur-train/2363737196

Mosasaurus
dkfindout.com/us/dinosaurs-and-prehistoric-life/prehistoric-reptiles/mosasaurus/

What Are Mosasaurs?
thekidshouldseethis.com/post/mosasaurs

INDEX